Miloš Poliak (Ed.)
Ivana Šimková

Transport Manager Responsibilities and Risk Based Insurance Analysis

Miloš Poliak (Ed.)
Ivana Šimková

Transport Manager Responsibilities and Risk Based Insurance Analysis

LAP LAMBERT Academic Publishing

Impressum / Imprint

Bibliografische Information der Deutschen Nationalbibliothek: Die Deutsche Nationalbibliothek verzeichnet diese Publikation in der Deutschen Nationalbibliografie; detaillierte bibliografische Daten sind im Internet über http://dnb.d-nb.de abrufbar.
Alle in diesem Buch genannten Marken und Produktnamen unterliegen warenzeichen-, marken- oder patentrechtlichem Schutz bzw. sind Warenzeichen oder eingetragene Warenzeichen der jeweiligen Inhaber. Die Wiedergabe von Marken, Produktnamen, Gebrauchsnamen, Handelsnamen, Warenbezeichnungen u.s.w. in diesem Werk berechtigt auch ohne besondere Kennzeichnung nicht zu der Annahme, dass solche Namen im Sinne der Warenzeichen- und Markenschutzgesetzgebung als frei zu betrachten wären und daher von jedermann benutzt werden dürften.

Bibliographic information published by the Deutsche Nationalbibliothek: The Deutsche Nationalbibliothek lists this publication in the Deutsche Nationalbibliografie; detailed bibliographic data are available in the Internet at http://dnb.d-nb.de.
Any brand names and product names mentioned in this book are subject to trademark, brand or patent protection and are trademarks or registered trademarks of their respective holders. The use of brand names, product names, common names, trade names, product descriptions etc. even without a particular marking in this works is in no way to be construed to mean that such names may be regarded as unrestricted in respect of trademark and brand protection legislation and could thus be used by anyone.

Coverbild / Cover image: www.ingimage.com

Verlag / Publisher:
LAP LAMBERT Academic Publishing
ist ein Imprint der / is a trademark of
OmniScriptum GmbH & Co. KG
Heinrich-Böcking-Str. 6-8, 66121 Saarbrücken, Deutschland / Germany
Email: info@lap-publishing.com

Herstellung: siehe letzte Seite /
Printed at: see last page
ISBN: 978-3-659-18214-3

Zugl. / Approved by: Žilina, University of Žilina, 2012

ABSTRACT

This thesis addresses the issue of the responsibility of the transport manager. The transport manager´s conditions that have to be respected during his/ her work are analyzed. The transport manager is responsible for the decisions made during the whole transportation process. Individual decisions are evaluated and a probabilistic estimate of their importance is determined. A methodology for insurance for the transport manager is proposed.

KEY WORDS: TRANSPORT MANAGER, INSURANCE, RESPONSIBILITY, DECISIONS

ABSTRAKT

V mojej práci sa budem venovať problematike zodpovednosti zasielateľa. Analyzovať zasielateľské podmienky, ktoré musí rešpektovať a dodržiadavať pri jeho práci. Zasielateľ je zodpovedný za celý proces prepravy a za jeho rozhodnutia ohladom prepravy. Budem hodnotiť rozhodnutia a určovať pravdepodobnosť výskytu a vyhodnocovať štatistické údaje. Navrhnem metodiku tvorby poistenia pre zasielateľa. Podľa dôležitosti rozhodnutí, ktoré sú dôležitejšie a ktoré by mali mať vyššie poistné krytie ako ostatné.

KĽÚČOVÉ SLOVÁ: ZASIELATEĽ, POISTENIE, ZODPOVEDNOSŤ, ROZHODNUTIA

TABLE OF CONTENTS

LIST OF TABLES

LIST OF FIGURES

CHAPTER 1: INTRODUCTION

1.1 BACKGROUND

The United States (U.S) has no insurance for the transport manager. Slovakia has an insurance for the transport manager, but some insurance companies do not know who the transport manager is and what are his/her responsibilities. This thesis is to review the conditions of insurance companies for the transport manager and their coverage.

The field of transportation management is a relatively new but very important and challenging field. The transport manager is an expert with knowledge concerning all kinds of vehicles and the transportation processes. He/she facilitates the transportation process and monitors shipments during the whole transportation process.

The transport manager plays a very important role for the general public involved in the transportation of goods. People may have difficulties to understand the transportation process involving goods, for example, which truck carrier is the best suited for a specific group of people or how the shipment would be processed in time and space. People are only concerned about their goods and their shipment from origin A to destination B in a safe and timely fashioned. This fact clearly indicates that people need the services of a transport manager.

The transport manager has to be very careful when making a decision during the whole transportation process, because a wrong decision could cause physical damage to the goods or delay the reception of the shipment.

For these reasons, the transport manager needs a really good insurance that covers the whole responsibility of the transportation process. The transport manager has to do his job without stress and without worries.

1.2 OBJECTIVES

The objectives of this thesis are: (1) to review the decisions the transport manager makes during the transportation process, (2) to evaluate and quantify the importance of

1

each decision, and (3) to compare a new proposed coverage with the existing coverage used by the insurance companies.

In addition, this thesis also analyses the difference between a transport manager in the U.S. and the transport manager in some European countries.

1.3 ORGANIZATION

Chapter 1 provides the background thesis organization, objectives and literature review. The literature review is limited, because this problem is kind new and nobody has studied it before.

Chapter 2 compares necessary conditions for starting a business in the field of transport manger between the U.S. and the European Union (EU). Transport manager must meet both the general and professional business conditions. The EU has general conditions that apply to the all EU members. But professional conditions differ from EU state to state.

The business conditions are less strict and friendlier in the US than in EU. When a transport manager wants to start a business in the US, he/she only needs registration for his or her own business. In addition, he/she needs a broker's license. The transport manager must have a special type of "insurance" known as a "Customs Bond". A bond is an insurance policy that guarantees payment to U.S. Customs and Border Protection (CBP) in case a required act is not performed.

Chapter 3 will discuss all the decisions of the transport manager. In that chapter, a diagram based on the time sequence of decisions of the transport manager will be proposed. An example scenario of the transportation process will be demonstrated through this diagram for a better understanding of the significance and the consequences of the decisions. The individual decisions that determine the probability of each event will be also evaluated. The probability that the transport manager will need an insurance coverage for the specific decision will be also estimated.

Chapter 4 describes the survey that was created to compare all of the decisions based on the frequency of the responses. Individual decisions are evaluated and used to determine the probability of each event. The probability that transport manager will need an insurance coverage for the specific decision is computed. Chapter 4 also describes statistical evaluation. Then, a methodology of insurance for the transport manager will be proposed. The methodology will focus on what kinds of decisions are the most important for the transport manager and which ones should be under the coverage of insurance is computed.

In Chapter 5, a comparison will be conducted between the proposed coverage and the coverage used by insurance company, as well as advantages and disadvantages will be drawn.

CHAPTER 2: ANALYSIS APPROACH TO ENTREPRENEURSHIP IN TRANSPORT MANAGEMENT

The definition of transport manager follows from the wording of particular paragraphs of the Commercial Code of Slovakia [2]. Paragraph 13 of this law contains all the provisions related to the transportation contract. According to that paragraph, the transport manager is the person that arranges the whole process of the transport and the customer is obliged to pay remuneration to him/ her.

2.1 ACTIVITIES OF THE TRANSPORT MANAGER IN THE EUROPEAN UNION (EU)

The transport manager has the responsibility to take professional care of the implementation of the responsibilities under the transportation contract. He/she is responsible for minimizing the transportation costs, minimizing risks arising from transportation and to ensure the best possible transportation of goods from source to destination. He/she represents the connection between the customers and the carrier, between the producers and the consumers, the exporters and the importers. His or her efforts are devoted to create the best price for customer. In this regard, he/she cooperates with motor carriers (transport capacity), and other transport managers. The transport manager also helps to speed up the flow of funds for goods sold by issuing documents accepted by the banks and the customs offices. In addition the transport manager is also an expert in various parameters concerning the vehicles, and he is able to recommend or select the appropriate vehicle for the transportation process. The transport manager combines shipments and modes of transport for a rational transport process. The transport manager is able to provide information to the customers on the movement of goods from the loading to the unloading site.

2.1.1 LEGISLATIVE CONDITIONS FOR MARKET ACCESS IN EU

The transport management across the EU has no regulations that would unite the conditions for transport manager. There is only a Directive of the European Parliament and Council 2006/123/ about services in the internal market [3]. This document governs only the education necessary for carrying out the business of the transport manager as follows:

- Secondary education and at least 2 years experiences in the field,

- A university degree and at least one year experience in the field.

For this reason, there are significant differences of approach in the forwarding business in the EU.

2.1.1.1 LEGISLATIVE CONDITIONS FOR MARKET ACCESS TRANSPORT MANAGER IN THE SLOVAK REPUBLIC

Business for transport management is regulated by law of services in the internal market [4]. According to these laws the transport manager must meet the following two criteria:

General business conditions:

- minimum age of 18 years,

- legal capacity,

- integrity.

Professional competence:

- secondary education and at least two years experiences in the field,

- university degree and at least one year experience in the field.

If the applicant for an authorization to operate in the transport management sector meets these requirements, the Trade Office is obliged to issue a business license.

In addition to the previous conditions of the law of services in the internal market [4], the transport manager has to fulfill another condition that will raise awareness to the customers. The customers will know more information about his/ her company and activities. Before signing a transportation contract or before providing any service, the transport manager should unambiguously inform the customer about:

- the business name, the legal status, the place of business or domicile, residence, the telephone number, fax number or address for electronic mail,
- identification number for value added tax, if assigned, otherwise the tax identification number, membership in professional organization, general conditions for the provision of forwarding services, liability if it provides a range of duties provided by law.
- cost of the service if the price is fixed in advance or paid, or method of calculating the price if the price will be set up to provide services, contact details of the insurer if he/she has contracted with the insurance, the extent of liability insurance for damage caused by execution of activities and spatial extent of the insurance, contact details, where the customer can lodge a complaint or claim to the service provided.

These data may be available to customers through:

- information available at the place where the service or where the contract is awarded for the provision of services,
- information documents intended to the recipient of services, which include a detailed description of the service,
- electronic means of communication, in particular the data at web site.

The Slovak Trade inspection has a control of compliance with an obligation under the law of services in the internal market [4].

2.1.1.2 LEGISLATIVE CONDITIONS FOR MARKET ACCESS TRANSPORT MANAGER IN AUSTRIA

The responsibilities of the transport manager are the same in Austria as in any other country. He/ she has to take professional care of the whole transportation process but the professional conditions are more strict than in other countries.

Professional education can also be obtained by successful accomplishing a test. Any individual who successfully completes the transport school plus at least 2 years of experience in the transportation management field is qualified to provide professional transportation services.

The topics of the test include:

1.   create correspondence and documents,

2.   payments and lending,

3.   calculate cost according to the tariffs,

4.   accounts under the special accounting,

5.   the oral examination.

2.1.1.3 LEGISLATIVE CONDITIONS FOR MARKET ACCESS TRANSPORT MANAGER IN THE CZECH REPUBLIC

The responsibilities of the transport manager are the same in the Czech Republic as in any other country, but the conditions are less strict than in other countries.

The transport manager must only meet the general conditions for business:

- minimum age 18 years old,

- legal capacity,

- integrity,

According to Government Resolution no. 140/2000 Collection of Law, transportation is divided into:

- national,

- international.

The transport manager cannot do transportation for his/her own purpose, representation in customs procedure and the activity of customs agents[5].

2.2.1 TRANSPORT MANAGER BUSINESS

The transport manager has to have really good knowledge of the transportation field. He/she needs knowledge of clearance process and of warehousing. That means that a transport manager can work as a manager of warehouse or an agent of the clearance process.

2.2.1.1 THE TRANSPORT MANAGER IN THE CLEARANCE PROCESS

The transport manager has to be familiar with conditions of clearance for each particular good during the transportation.

Transport manager as agent of the declarer

Under the Customs Act [7], the declarer (the consignor - the person responsible for the customs debt) may designate a representative as:

- direct representation - the representative acts in his own behalf and takes all the risks,
- indirect representation - the representative acts on his own behalf but the risks are taken by the represented person.

The representation must be ensured by one of following:

- an authorization contract,
- a commission contract – in case of the indirect representation,
- a mandate contract – in case of the direct representation.

Documents required for representation in customs procedure

If the transport manager wants to represent the client in the customs procedure he/she must have these papers:

- proof of representation; it is contract between the transport manager and the customer about the representation in customs procedure,

8

- extract from the Commercial Register of the declarer (the transport manager); like a proof that the transport manager really exists,
- tax Office decision to grant a tax registration number of the declarer; the transport manager needs this number for the business, without this number he/she cannot do his or her job.
- proof of security of customs debt:
  - contract between customs office and transport manager about transfer cash to the customs office,
  - the insurance contract between the insurance company and the transport manager to cover his or her responsibility for the customs procedure.

Procedure to obtain authorization to act as guarantor for customs debts:

1) provision of guarantees to secure the customs debt (business license, certificate of incorporation),
2) calculate the estimated amount of the customs debt and the customs office for its approval and adoption,
3) to contract an insurance with the insurance institute, which must specify the subject matter of the insurance - liability for damage caused by the shipper's failure to guarantee the customs debt,
4) fill in the warranty deed to the competent customs office,
5) the customs office, after examining the application shall issue a decision which allows the transport manager to provide guarantees.

After all these steps he/she can work as guarantor for customs debts.

Transport manager as a warehouse operator

Customs warehouse facility is a secured area where goods are stored under the conditions stipulated by the customs office. Customs warehouse may be operated as a public or non-public warehouse.

Public warehouse

Public warehouse can be used for storage of goods of any customers. Goods in the warehouse are stored for remuneration. Public warehouse may operate only under permit issued by the locally competent customs office in whose jurisdiction the warehouse is located.

Non-public (private) warehouse

Non-public warehouse is the same as public warehouse, but the goods in the warehouse are stored without fees. This warehouse is only for members of the customs office.

The goods in the warehouse cannot be released to the market when:
- the import duty and tax for goods are not paid
- the customs documents are not complete
- the recipient is not in the country where the warehouse is located
- the goods are damaged
- there are doubts about the origin of goods

2.2.1.2 THE WAREHOUSING BUSINESS

Warehousing is short or long-term storage of assets, raw materials, intermediate products, finished products in the own warehouse or in equipment and premises of another person.

The transport manager does these services in storage of goods in warehousing:
- evidence of goods in the storage,
- sorting of goods,
- preparation before transport,
- the labeling,
- repackaging,
- consolidation and deconsolidation of the goods.

The transport manager prepares the goods in the order of how they are needed for manufacturing. Parts must be transported on line just in time, but also in the correct sequence.

Legislation

Pursuant to the law 455/1991 Collection of laws [8], Storage is classified as free trade, but does provide other ancillary services (handling of consignment, packing, etc.)

The contract of storage is governed by the Commercial Code (law 513/2001 Collection of laws), which recognizes the following contracts the storage contract that governs the contractual relationship between the provider of storage services and its customer (depositor, customer)

The storage contract should contain:

- Definition of Parties
- Subject of Contract
- Commitments
- The price of storage
- The rights and obligations of the transport manager and customer
- The date of payment for services rendered
- Validity and termination of the contract

The storage contract (paid storage)

The warehouse undertakes the items intended for storage and provides care; the depositor agrees to pay for this service.

The storage contract (storage for free)

The service provided by this contract is operated under the same conditions as the previous storage contract (paid storage), but this is free of charge in some special cases such as regular customer or high amount of goods stored.

The advantages if the transport manager provides of storage services

- The goods stored with the transport manager are available for customers.
- The company has accurate information on stock status (on-line storage systems)
- If the transport manager is working under the general transport manager's conditions, these conditions define rights and obligations of the transport manager for the storing goods.

Rights and obligations of the transport manager under the general transport management conditions

The goods are stored in a warehouse by the transport manager of choice (for the location of goods in a foreign stock the transport manager is obliged to inform the customer about the name of the storekeeper, and location).

The customer has the right to inspect the storage space and has an objection to the quality of storage. If the customer, immediately after inspecting the warehouse has no objection, he/she waives the possibility of subsequent complaints on quality of warehouse.

The customer has the right to enter the warehouse only when accompanied by the transport manager or the storekeeper. The customer is liable for all damages caused to the warehouse when the goods is stored by others.

2.2 THE TRANSPORT MANAGER AND HIS ACTIVITIES IN THE U.S.

Transportation manager is employed by companies that require the coordination of different means of transportation. A transportation manager may work for a number of different companies. He may work for a bus company that provides tourist busing and tours and services. He may work for a school bus company that provides transportation to and from school for children. He may also work for a truck driving company that provides truck transporting of goods and materials for outside clients.

The duties of a transportation manager can vary depending on the company he works for. For example, a manager who works for an independent trucking company hired by clients may have to be constantly designing new routes depending on demand for transportation. Likewise, a transportation manager who works for a large moving company will have to design new routes depending on clients who call to move. This may involve finding a driver willing to drive that particular route, allocating a vehicle from the fleet to that route, setting a price for the route, mapping the route, and taking care of other contingencies associated with the route itself.

In other situations, the routes the trucks or vehicles take may be the same on a daily or weekly basis. For example, a school bus company generally takes the same route to and from school every day, picking up the same children. In such cases, the route planning done by the manager would be done at the beginning of the school year and modified only if new students come to the school or if other changes must be made. The transportation manager in this type of company would thus have a different set of responsibilities, such as ensuring all drivers comply with state and local laws for child safety and that all vehicles are properly maintained, registrations are up-to-date, and that safety standards and codes are met on the vehicles.

Often, managers involved in transportation learn many of the skills they need on the job, and must be very organized, have strong attention to detail and have good people skills to interact with both clients and drivers. A college degree may or may not be required, depending on the company and the extent of the fleet being managed. Transportation, hospitality or management degrees may all be sufficient education to get into the field of transportation management[8].

2.2.1 ACTIVITIES OF THE TRANSPORT MANAGER

The work activities of transport managers vary significantly depending on the industry or sector and the size of the organization. Senior posts encompass strategic planning and project management work, while junior posts are more involved in customer contact and routine staff supervision. In general, however, typical work activities include:

- making sure the operation meets its performance and safety targets, monitoring and reporting performance to senior management,
- writing clear reports and presenting options and recommendations to clients and senior management,
- advising on policy and strategic developments, examining business decisions (pricing policy, level of service provision, timetable changes) to assess their impact on passengers,
- ensuring that all operations are carried out in accordance with UK and European Union (EU) laws and regulations, particularly relating to health and safety,
- managing and supervising staff, organizing work shift rotes and coordinating staff training,
- negotiating and managing contracts, developing new business opportunities whenever possible,
- minimizing disruption and resolving any unscheduled delays, making decisions in difficult situations,
- meeting passengers and customers, dealing with complaints and areas of concern;
- analyzing results of surveys on passenger/customer satisfaction, instigating new projects to improve performance,
- ensuring that transport services are available to all through social inclusion initiatives;
- marketing passenger services to encourage greater passenger use of particular routes and methods of transport,

- liaising and negotiating with different stakeholders including planning and highways authorities, residents, councilors/ politicians, developers and transport providers,
- identifying existing and possible future transport problems, developing transport models and investigating the feasibility of alternative means of transport,
- liaising with passenger watchdogs and other professional bodies,
- using IT systems for tasks such as timetabling and managing usage flows[8].

2.2.2 LEGISLATIVE CONDITIONS FOR MARKET ACCESS IN U.S.

The transport manager can set up business as a "sole proprietor" (single person). To do so, the transport manager just starts his own business.

There are three main types of corporate forms for businesses - "C" corporation, "S" corporation, and LLC or Limited Liability Company. Most small businesses choose to be either an "S" corporation or an LLC because these forms limit personal liability while still allowing profits and losses to be passed through to the owners' personal tax returns[10]. Every transport manager must be a licensed customs broker, which requires passing an exam.

To be eligible for a broker´s license:
- be at least 21 years old,
- be a United States citizen,
- not be a current Federal Government employee,
- possess good moral character.

If the transport manager complies with previous requirements, he/she additionally has to:
- pass the customs broker license examination,
- submit a broker license application with appropriate fees,

- application must be approved by CBP (U.S. Customs and Border Protection).

The customs Broker license examination is an open book/open test with 80 multiple-choice questions based on designated editions of the following publications:

- the Harmonized Tariff Schedule of the United States (HTSUS),
- title 19, Code of Federal Regulations,
- specified Customs Directives,
- customs and Trade Automated Interface Requirements document (CATAIR).

The appropriate CBP port director must receive the examination application and a $200 fee at least 30 days in advance of the examination. An acceptable result is considered when the applicant gets at least a 75 % score[11].

The transport manager must have a special type of "insurance" called a customs bond.

A CBP bond is a contract that is given to insure the performance of an obligation or obligations imposed by law or regulation. A bond is like an insurance policy that guarantees payment to U.S. Customs and Border Protection (CBP) if a required act is not performed. Bonds have a number of uses in CBP. The most common use allows importers to take possession of their goods before all CBP formalities are completed. Another common use allows a carrier to move goods under bond from one place to another before those goods are actually entered for consumption with duties paid.

All parties that import merchandise into the United States for commercial purposes or transport imported merchandise through the United States must have a CBP Bond[12].

2.3 Comparison of legislative conditions for market access in U.S. and EU

For a transport manager it is easier to start a new business in the U.S. than in some countries in EU. But conditions are same all over the U.S. In the U.S. there is just one condition that the transport manager must have: the customs Broker license. The transport manager obtains the customs Broker license when he passes a test.

The EU has just general conditions, which apply to the whole EU. However professional conditions are different in each country. Some countries have less strict conditions (Czech Republic) for transport managers and some countries have stricter ones (Austria). The EU needs to unify professional conditions for all the EU state members; similarly as it is in the general business because the competitive environment is distorted.

The transport manager in U.S. must have a special type of "insurance" called a customs bond, in the case transportation is from one country to another. The European transport manager does not need this insurance when transport is within European union (27 countries), but the transport manager must have this type of insurance when transportation is performed from EU to another country.

CHAPTER 3: ANALYSIS OF TRANSPORT MANGER´S LIABILITY

This chapter focuses on the responsibility of a transport manager in Slovakia. The transport manager´s responsibility is adjusted by law §601- 609 [2] and by transport manager´s conditions[13]. The transport manager´s conditions specify transport manager´s responsibility. If transport manager does not use these conditions, he/she is responsible for all damage. The conditions must to be an annex of the transportation contract.

Since the transport manager is responsible for the whole transportation process and for his or her decisions of the transportation process. He/she is responsible for following decisions:

- ordering transport timely,
- knowing correct information about goods,
- neglect to provide professional care at order of transportation,
- responsibility for wrong decisions,
- correct handling of goods,
- risk of storage of goods during transportation,
- responsibility for averting damage to the goods,
- responsibility for another transport manager (when he/she use another transport   manager),
- neglecting of sell of goods, if goods can be destroyed,
- non- application of his/her rights,
- not to inform customer about destruction of the goods,
- notice to the wrong customer´s instructions about transportation,
- responsibility for the completion of the instructions,
- insurance of the goods,
- filling the bill of lading,
- late delivery,
- correct vehicle,

- price,
- dates of transportation.

3.1 THE TRANSPORT MANAGER'S RESPONSIBILITY

Transport managers in Slovakia use the transport manager's conditions stipulated by the Association of Logistic and Forwarding of Slovakia. If transport manager has applied these conditions, they have to be attached to the contract of transportation. These conditions are used to protect both the transport manager and the customer. That means the transport manager knows what his/her duties and rights are, and the customer knows what to expect from transport manager.

The transport manager's conditions regulate:

- definition of contracting parties. Transport manager's and customer's name, full address, mail, phone number,
- conclusion of the contract of transport. The form of contract between customer and transport manager about transportation,
- conditions of admission of the order of transport. What order form can be accepted for the transportation process,
- rights and obligations for customer and transport manager. What things the transport manager and customer have to do during transportation,
- customer's and transport manager's responsibility of the damage of transport. What are the transport manager and customer responsible for during transportation,
- grounds for exemption from responsibility of the damage of transport. In some cases transport manager is not responsible for damage. For example he/she cannot be responsible for natural disaster, because he/she has no control over it,
- conditions for storage. How do goods have to be stored, what conditions does transport manager have to apply for transportation,

- invoicing and payment terms of transport contract. As a customer can pay for transportation to the transport manager and how many days has of pay.
- contract, penalty and right of retention. The form of contract, penalty in the case that customer did not pay for transport and right to not to issue goods if the customer refuses to pay

The transport manager's responsibility for damage by the transport manager's conditions

The damage which occurred during transport or handling is limited:
- 8,33 XDR (Special Drawings Rights) per 1 kilogram weighted, damaged, destroyed or lost goods, but
- maximum 20 000 XDR for whole transportation process.

  In the case that goods are not delivery to place of delivery timely, his/her responsibility is limited to the cost of transport.

The damage, which occurred during storage, is limited:
- 3, 925 XDR per 1 kilogram, but
- maximum 3 925 XDR for whole transportation process.

In case of other damages, for example, if he/she does something wrong during transportation process or makes a wrong decision his/her responsibility is limited to 20 000 XDR for whole transportation process.

The transport manager and customer can change the transport manager's conditions for some special transportation. If the goods are of high value, then the responsibility is high too. If the transport manager caused the damage intentionally he/she paid in full for the damage.

Cases when the transport manager is not responsible for damage include:
- the damage that occurred independently of the transport manager,
- if customer does not accept the advice of the transport manager,

- if the transport manager cannot prevent some obstacles or effects,
- if the transport manager could not know about some obstacles or effects.

The transport manager is excused from responsibility in the case of floods, fires, hurricanes, tsunamis, tornadoes, and calamities.

If the transport manager does not use these conditions, his/her responsibility is unlimited.

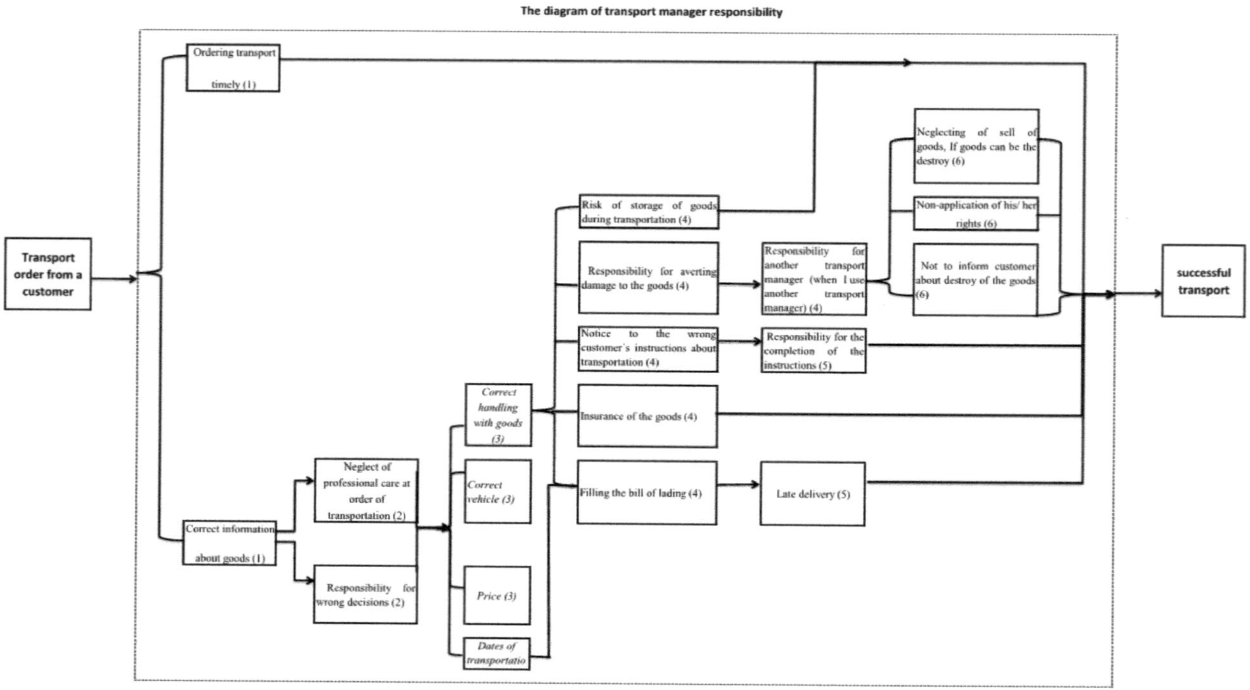

The diagram of transport manager responsibility

Figure 3.1: The diagram of the transport manager responsibility

22

3.2 EXAMPLE OF TRANSPORT MANAGER RESPONSIBILITY (DIAGRAM)

To get a better insight of all responsibilities of the transport manager, a fictional transportation will be created to simulate what could happen for each decision of transport manager if he/she makes a wrong decision during transport. Let the transport take place between 15th of January to 20th of January. The last day for forwarding is 20th of January. The goods are 2880 pieces of ice creams (Figure 3.2). The transport is from Žilina (Slovakia) to Paris (France). The size of piece of the ice cream is 12x 18x 6 centimeters and the weight is 560 grams.

Figure 3.2: Box of the ice cream

Solution:

Figure 3.3 Route that the vehicle will follow between Žilina and Paris.

Destination B= Paris   Total distance is 1 438 km   Origin A= Žilina

The goods will be transported in cardboard boxes. The capacity of each cardboard box is 90 pieces of ice creams. The capacity of EURO pallet is 720 pieces of ice creams. 4 EURO pallets are needed to transport the ice cream. The total weight of the parcel is 161.28 kg. Figure 3.4 and 3.5 show the overall dimensions of the whole parcel.

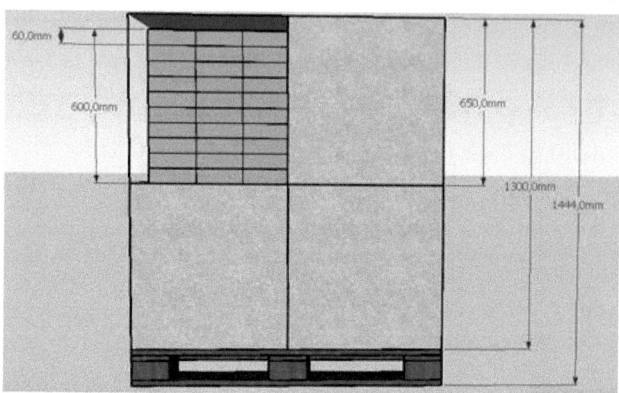

Figure 3.4: The goods on the EURO pallet from the front side

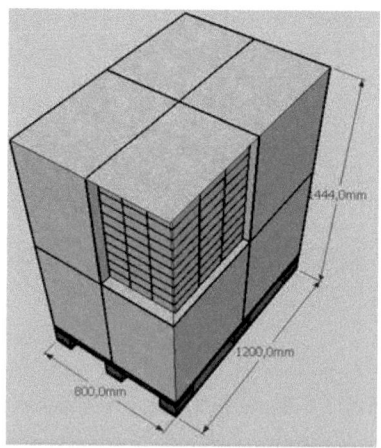

Figure 3.5: The goods on the EURO pallet from the side

Ice cream is considered a special good and the transport manager must know the conditions for transportation. The conditions are in agreement with ATP (Agreement on the International Carriage of Perishable Foodstuffs) [http://www.unece.org/trans/main/wp11/atp.html]. The ice creams must be transported in a vehicle with machine- cooling device, which must allow choosing the temperature inside the cabinet between + 12 ° C to - 20 ° C.

In this case the transport manager uses a Mercedes- Benz SPRINTER 315 2.2 CDI truck. The dimensions of the vehicle are 255x 186x 170 centimeters (4x EURO pallets) with a machine- cooling device.

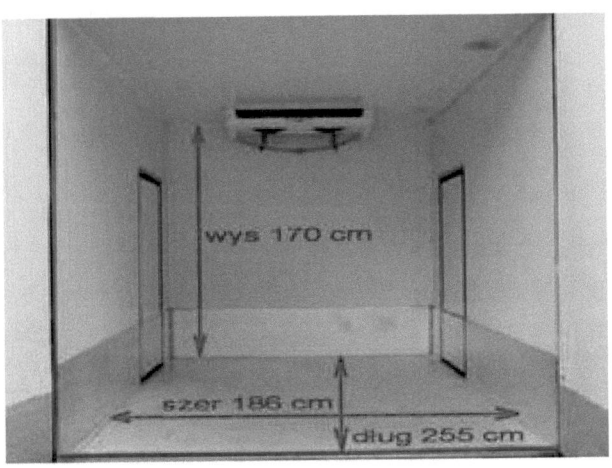

Figure 3.6: Inside of the vehicle

Figure 3.7: The vehicle from the outside

The transport manager's responsibility for transportation

Next paragraphs describe each of the decisions that the transport manager has to make in the transportation process. Each decision has a number from 1 to 6 associated to it. These numbers represent the importance of the decision. The number 1 is the highest value for the decision and the number 6 is the lowest value for the decision.

Ordering transport timely (1)

At least one day before loading goods

Correct information about goods (1)

The transport manager must have the correct information about goods and correctly give these information to carrier. The transportation is from Žilina (Slovakia) to Paris (France), goods are ice creams on the 4 EURO pallets. The transport manager must use a vehicle with machine- cooling device, date of loading is 1/15/2012 and date of delivery is 1/16/2012.

Neglect of professional care at order of transportation (2)

The transport manager must know the correct conditions about transportation of ice creams, which vehicle and packaging is appropriate for transporting the ice cream. The transport manager must know the agreement of ATP. Ice creams will be transported in cardboard boxes on the EURO pallets see (Figure 3.4 and Figure 3.5).

Responsibility for wrong decisions (2)

The transport manager must make the correct decisions of transport process:

- If he/she chooses wrong vehicle without machine- cooling device, ice cream will be destroyed,
- If he/she chooses wrong packaging, packaging of ice creams will be destroyed,
- If he/she orders transportation too late, goods will not be in the place at the right time.

Correct handling with goods (3)

The transport manager must specify to the carrier, that goods are ice creams. Ice cream must be transported at $-20\ °C$

Risk of storage of goods during transportation (4)

If necessary, obtain suitable warehouse with deep-freezer for the best price.

Responsibility for averting damage to the goods (4)

The transport manager must make the decision, because he/she must protect the goods to avoid that they get destroyed. In the case that the machine- cooling device is broken during transportation, transport manager must a find new vehicle with a similar type of device, before the ice cream melts.

Responsibility for another transport manager (when transport manager uses another transport manager) (4)

The transport manager is responsible for the success of the transport. That means, he/she is responsible for all persons in the transportation process.

Neglecting the sell of goods, if goods can be the destroyed (6)

For example, some natural disaster could damage the roads and the transportation could not continue. The cost for transportation would increase more than the customer paid for the transportation. The transport manager must tell about this situation to the customer. If the customer does not have any suggestions, any ideas, the transport manager recommends the sell of the goods. Also he is responsible for the best price for the goods.

Non-application of his/her rights (6)

The transport manager must ask his/her payment for successful transportation. Because after three years his/her claim on the payment expires.

Not to inform customer about damage of the goods (6)

The transport manager cannot make any decision without letting the customer know. He/she must put information to customer about any situation. In case that customer has no answers, the transport manager makes the decisions. He/she makes the best decision for customer and when the customer is available, he/she informs the customer about the situation.

Notice to the wrong customer's instructions about transportation (4)

The transport manager must check all instructions about transportation and whether these instructions are correct. In case any instructions are wrong, he/she must tell this information to customer and these instructions must be corrected.

Responsibility for the completion of the instructions (5)

The transport manager must ask customer about completion of the instructions of the transportation in a formal communication (letter or mail).

Insurance of the goods (4)

Sometimes transport manager needs additional insurance for the goods being transported when goods are expensive in case something should happen.

Filling the bill of lading (4)

The correct information of the goods (e.g. address, phone number of carrier, delivery time, etc.)

Late delivery (5)

The transport manager must use all his/her ability for timely deliver the goods to place of delivery.

Correct vehicle (3)

The transport manager must use vehicle mode F (FRIGO) with machine- cooling device and with certificate ATP.

Price (3)

After the transport manager and customer have agreed on the price of transporting the goods, the transport manager cannot change the price. The price is calculated by number of kilometers x price for kilometers + 10% fee for the transport manager for successful transportation.

Dates of transportation (3)

The loading is done on 1/15/2012 and delivery is made on 1/20/2012 at the latest.

The transport manager has a big responsibility in the whole transportation process. For transport manager it is important to make a proper decision. Each decision has different consequences, but every single one could endanger the whole transportation process.

3.3 CLASSIFICATION OF DECISIONS

The classification of the decisions is needed to assess the importance of the decisions and to calculate the coverage of the insurance. A survey was prepared as a means to identify the most important decision transport managers make in the transport process. The survey includes all the decisions (19 decisions), which a transport manager has to make during a typical transportation task. Respondents of the survey have to assign a value to each decision in a scale of it is to define the importance of each decision. In this scale one means not important and five means most important.

The values (points) from one to five were chosen, because more values mean a large variety of answers. There are five possible values that mean each of the values represent 20% (100% /5= 20%). A value of one means that the decision is not important (20%), two is less important (40%), three is important (60%), four is very important (80%) and five is most important (100%).

The survey was sent to six companies (SETTO spedition s.r.o., Vadual logistik spol. s r.o., ŠIKOTRANS, s.r.o. Kovotvar v.d., SPK International s.r.o. and COLSPEDIA, s.r.o.). These companies were chosen because the autor worked at three of these companies (SETTO spedition s.r.o., ŠIKOTRANS s.r.o., SPK International s.r.o.) and have really good relationships with their employees. The other companies were chosen because of the existing good connection with them. A total of 48 persons returned the completed survey.

The results from the survey are graphically illustrated to show the frequency of the answers in figures 3.7 to 3.25. Each figure shows a pie chart with the percentage frequency of the importance assigned to a particular decision by the respondents. The corresponding frequency table is shown at the left side of the figure. The largest number of responses is highlighted in bold numbers.

- Decision 1: Ordering transport timely

	1 (0%)
	2 (12.5%)
	3 (29.2%)
	4 (33.4%)
	5 (25%)

Points	Answers
1	0
2	6
3	14
4	**16**
5	12

Figure 3.8: Relative frequency of survey responses and frequency for decision 1.

A total of 16 people assign the value 4 (80 %) to importance of this decision

- Decision 2: Correct information about goods

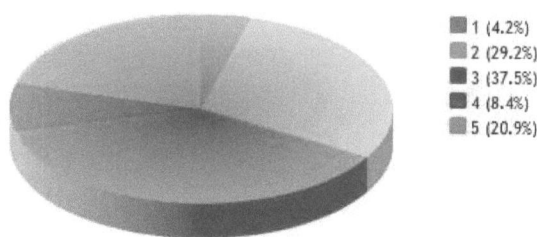

1 (4.2%)
2 (29.2%)
3 (37.5%)
4 (8.4%)
5 (20.9%)

Points	Answers
1	2
2	14
3	18
4	4
5	10

Figure 3.9: Relative frequency of survey responses and frequency for decision 2

A total of 18 people assign the value 3 (60 %) to importance of this decision.

- Decision 3: Correct vehicle

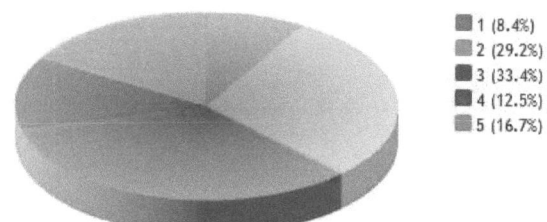

1 (8.4%)
2 (29.2%)
3 (33.4%)
4 (12.5%)
5 (16.7%)

Points	Answers
1	4
2	14
3	18
4	4
5	10

Figure 3.10: Relative frequency of survey responses and frequency for decision 3

A total of 18 people assign the value 3 (60 %) to importance of this decision.

- Decision 4: Dates of transportation

1 (0%)
2 (20.9%)
3 (16.7%)
4 (33.4%)
5 (29.2%)

Points	Answers
1	0
2	10
3	8
4	**16**
5	14

Figure 3.11: Relative frequency of survey responses and frequency for decision 4

A total of 16 people assign the value 4 (80 %) to importance of this decision.

- Decision 5: Price

1 (4.2%)
2 (33.4%)
3 (33.4%)
4 (8.4%)
5 (20.9%)

Points	Answers
1	2
2	**16**
3	**16**
4	4
5	10

Figure 3.12: Relative frequency of survey responses and frequency for decision 5

A total of 16 people assign the value 2 (40 %) and the value 3 (60 %) to importance of this decision.

- Decision 6: Notice to the wrong customer's instructions about transportation

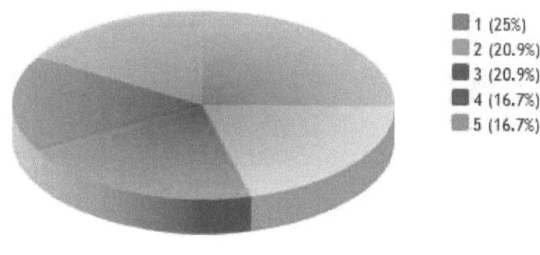

1 (25%)		Points	Answers
2 (20.9%)		**1**	**12**
3 (20.9%)		2	10
4 (16.7%)		3	10
5 (16.7%)		4	8
		5	8

Figure 3.13: Relative frequency of survey responses and frequency for decision 6

A total of 12 people assign the value 1 (20 %) to importance of this decision.

- Decision 7: Responsibility for the completion of the instructions

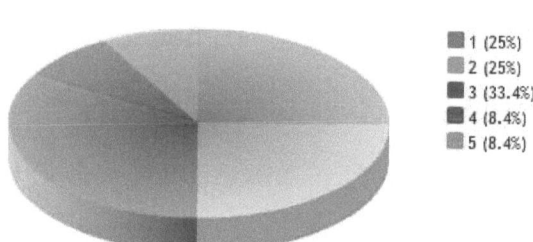

1 (25%)		Points	Answers
2 (25%)		1	12
3 (33.4%)		2	12
4 (8.4%)		**3**	**16**
5 (8.4%)		4	4
		5	4

Figure 3.14: Relative frequency of survey responses and frequency for decision 7

A total of 16 people assign the value 3 (60 %) to importance of this decision.

- Decision 8: Responsibility for wrong decisions

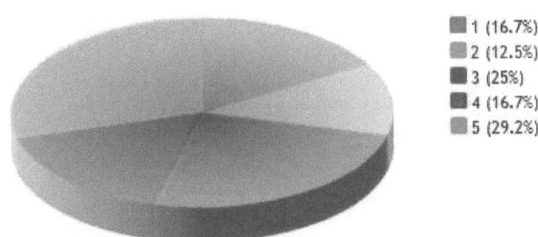

Points	Answers
1	8
2	6
3	12
4	8
5	**14**

1 (16.7%)
2 (12.5%)
3 (25%)
4 (16.7%)
5 (29.2%)

Figure 3.15: Relative frequency of survey responses and frequency for decision 8

A total of 14 people assign the value 5 (100 %) to importance of this decision.

- Decision 9: Late delivery

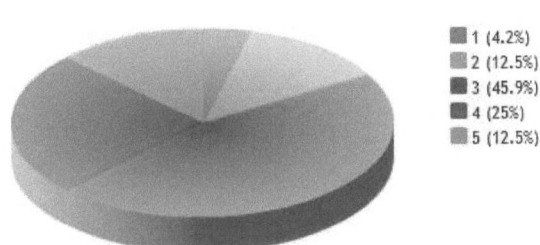

Points	Answers
1	2
2	6
3	**22**
4	12
5	6

1 (4.2%)
2 (12.5%)
3 (45.9%)
4 (25%)
5 (12.5%)

Figure 3.16: Relative frequency of survey responses and frequency for decision 9

A total of 22 people assign the value 3 (60 %) to importance of this decision.

- Decision 10: Correct handling with goods

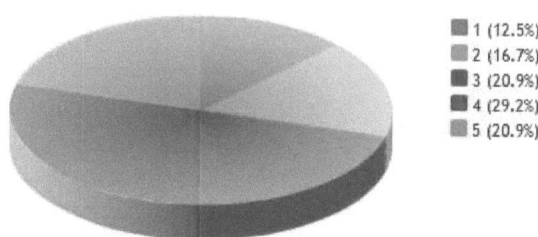

1 (12.5%)
2 (16.7%)
3 (20.9%)
4 (29.2%)
5 (20.9%)

Points	Answers
1	6
2	8
3	10
4	14
5	10

Figure 3.17: Relative frequency of survey responses and frequency for decision 10

A total of 14 people assign the value 4 (80 %) to importance of this decision.

- Decision 11: Neglect of professional care at order of transportation

1 (8.4%)
2 (12.5%)
3 (33.4%)
4 (25%)
5 (20.9%)

Points	Answers
1	4
2	6
3	16
4	12
5	10

Figure 3.18: Relative frequency of survey responses and frequency for decision 11

A total of 16 people assign the value 3 (60 %) to importance of this decision.

- Decision 12: Not to inform customer about destroy of the goods

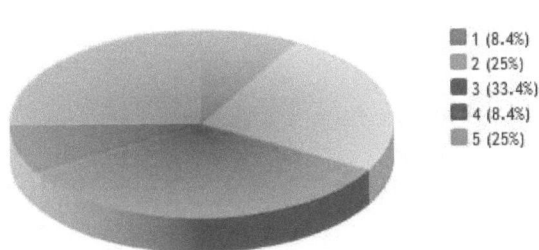

1 (8.4%)
2 (25%)
3 (33.4%)
4 (8.4%)
5 (25%)

Points	Answers
1	4
2	12
3	**16**
4	4
5	12

Figure 3.19: Relative frequency of survey responses and frequency for decision 12

A total of 16 people assign the value 3 (60 %) to importance of this decision.

- Decision 13: Insurance of the goods

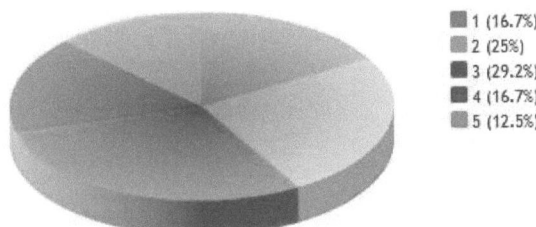

1 (16.7%)
2 (25%)
3 (29.2%)
4 (16.7%)
5 (12.5%)

Points	Answers
1	8
2	12
3	**14**
4	8
5	6

Figure 3.20: Relative frequency of survey responses and frequency for decision 13

A total of 14 people assign the value 3 (60 %) to importance of this decision.

- Decision 14: Responsibility for another transport manager (when transport manager uses another transport manager)

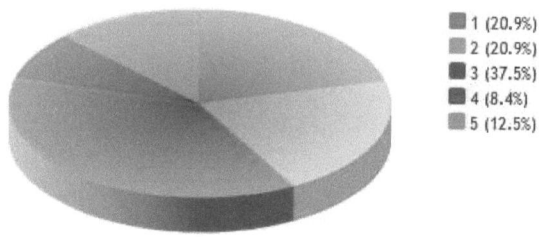

1 (20.9%)
2 (20.9%)
3 (37.5%)
4 (8.4%)
5 (12.5%)

Points	Answers
1	10
2	10
3	18
4	4
5	6

Figure 3.21: Relative frequency of survey responses and frequency for decision 14

A total of 18 people assign the value 3 (60 %) to importance of this decision.

- Decision 15: Responsibility for averting damage to the goods

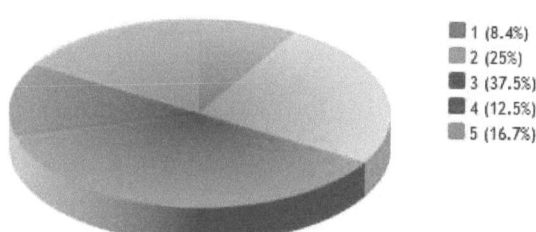

1 (8.4%)
2 (25%)
3 (37.5%)
4 (12.5%)
5 (16.7%)

Points	Answers
1	4
2	12
3	18
4	6
5	8

Figure 3.22: Relative frequency of survey responses and frequency for decision 15

A total of 18 people assign the value 3 (60 %) to importance of this decision.

- Decision 16: Neglecting of sell of goods, If goods can be the destroy

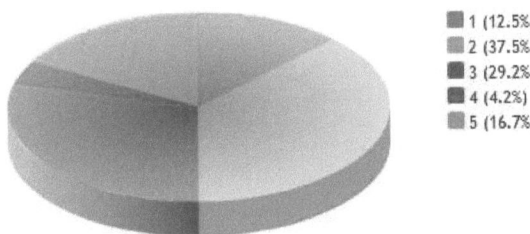

1 (12.5%)
2 (37.5%)
3 (29.2%)
4 (4.2%)
5 (16.7%)

Points	Answers
1	6
2	18
3	14
4	2
5	8

Figure 3.23: Relative frequency of survey responses and frequency for decision 16

A total of 18 people assign the value 2 (40 %) to importance of this decision.

- Decision 17: Non-application of his or her rights of the goods

1 (25%)
2 (29.2%)
3 (12.5%)
4 (20.9%)
5 (12.5%)

Points	Answers
1	12
2	14
3	6
4	10
5	6

Figure 3.24: Relative frequency of survey responses and frequency for decision 17

A total of 14 people assign the value 2 (40 %) to importance of this decision.

- Decision 18: Filling the bill of lading

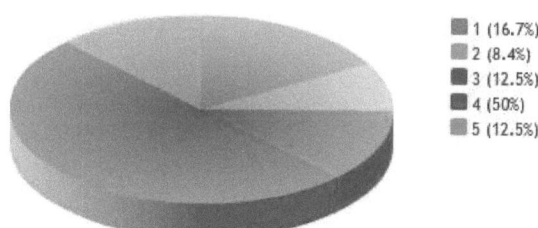

Points	Answers
1	8
2	4
3	6
4	**24**
5	6

Legend: 1 (16.7%), 2 (8.4%), 3 (12.5%), 4 (50%), 5 (12.5%)

Figure 3.25: Relative frequency of survey responses and frequency for decision 18

A total of 24 people assign the value 4 (80 %) to importance of this decision.

- Decision 19: Risk of storage of goods during transportation

Points	Answers
1	6
2	4
3	**14**
4	10
5	**14**

Legend: 1 (12.5%), 2 (8.4%), 3 (29.2%), 4 (20.9%), 5 (29.2%)

Figure 3.26: Relative frequency of survey responses and frequency for decision 19

A total of 14 people assign the value 3 (60 %) and the value 5 (100 %) to importance of this decision.

CHAPTER 4: METHODOLOGY FOR DEVELOPMENT OF INSURANCE OF TRANSPORT MANAGER

In this chapter the results from the survey discussed in Chapter 3 are proposed to rank decisions according to their importance level. The insurance methodology that is proposed in this chapter uses the ranking of the decisions as an input.

4.1. FORMULAS OF HOW TO FIND VALUE OF THE DECISION

Two formulas to assign the importance value of a decision are proposed. The first formula is the weighted average of the responses to each answer normalized with respect to the maximum decision value (5).

$$Q_S = \frac{\sum_{i=1}^{5} P_i \times Q_{pi}}{5 \; x \; n} \tag{4.1}$$

Where:

P_i is question importance i

Q_{pi} is frequency of response i

n is number of people who returned the survey

The second formula gives the importance value of a decision with respect to the entire survey. This formula was proposed, because it is needed to know the influence in percent of the each decision of the transport process with respect to the entire survey.

$$Q_n = \frac{Q_s}{\overline{Q_s}}$$

(4.2)

Where:

Q_s weighted average of the responses

$\overline{Q_s}$ is sum of the results from first formula

Example of using Formula 4.1 and the survey results for decision 1 shown in Table 4.1

Table 4.1: Example of using Formula 4.1

Question	Points	Answers
	1	0
	2	6
Ordering transport timely	3	14
	4	16
	5	12
Sum	-----	48

$$Q_s = \frac{1x\ 0 + 2x\ 6 + 3x\ 14 + 4x\ 16 + 5x\ 12}{5x\ 48} = 0.74166$$

The weighted average of question one with respect to only question one is 0.74166 (74.16 %). The highest possible average of the each question is one (100%).

42

Results for all the other decisions are shown in Table 4.2. Table 4.3 shows the results ordered in descending order.

Table 4.2: Tables of results of Formula 4.1

Formulas 4.1 sorted

Table 4.3: Tables of results of

Q_s		
Question	Value	Percent
1	0.74166	74.17%
2	0.625	62.50%
3	0.6	60.00%
4	0.74166	74.17%
5	0.61666	61.67%
6	0.55833	55.83%
7	0.5	50.00%
8	0.65833	65.83%
9	0.65833	65.83%
10	0.65833	65.83%
11	0.675	67.50%
12	0.63333	63.33%
13	0.56666	56.67%
14	0.54166	54.17%
15	0.60833	60.83%
16	0.55	55.00%
17	0.53333	53.33%
18	0.66666	66.67%
19	0.69166	69.17%
SUM	11.82493	-------

Ranked from higher to lower	
Q_s	
Question	Percent
4	74.17%
1	74.17%
19	69.17%
11	67.50%
18	66.67%
10	65.83%
9	65.83%
8	65.83%
12	63.33%
2	62.50%
5	61.67%
15	60.83%
3	60.00%
13	56.67%
6	55.83%
16	55.00%
14	54.17%
17	53.33%
7	50.00%

The most important decisions according to Formula 4.1 are decision number four and decision number one with an importance value of 0.74166 (74.17%).

Decision four is the date of transportation. It is very important to follow the date of loading and delivery for a successful transportation process.

Decision one is to order transport timely. This decision is also important; a transport manager has to order transportation timely. If he/she does not order transport timely the goods will be delivered later. The consequences of these decisions are described in section 3.2.

When Formula 4.2 is used, results are shown in Table 4.4. Table 4.5 shows the results sorted

$$Q_n = \frac{0.74166}{11.82493} = 0.06272$$

The value for question one with respect to entire questionnaire is 0.06272 (6.27 %). This number defines the influence of each decision of the transport process with respect to the entire questionnaire.

Table 4.4: Tables of results of Formula 4.2 Formulas 4.2/sorted

Q_n		
Question	Value	Percent
1	0.06272	6.27%
2	0.052854	5.29%
3	0.049331	4.93%
4	0.06272	6.27%
5	0.052149	5.21%
6	0.047216	4.72%
7	0.042283	4.23%
8	0.055673	5.57%
9	0.055673	5.57%
10	0.055673	5.57%
11	0.057082	5.71%
12	0.053559	5.36%
13	0.047921	4.79%
14	0.045807	4.58%
15	0.051445	5.14%
16	0.046512	4.65%
17	0.045102	4.51%
18	0.056378	5.64%
19	0.058492	5.85%
SUM	1	100%

Table 4.5: Tables of results of

Ranked from higher to lower Q_n	
Question	Percent
1	6.27%
4	6.27%
19	5.85%
11	5.71%
18	5.64%
8	5.57%
9	5.57%
10	5.57%
12	5.36%
2	5.29%
5	5.21%
15	5.14%
3	5.07%
13	4.79%
6	4.72%
16	4.65%
14	4.58%
17	4.51%
7	4.23%
SUM	100%

The most important decisions according with Formula 4.2 are decision number four and decision number one with value 0.0627 (6.27 %). The result is the same as using Formula 4.1, but this value is with respect to the entire questionnaire. However, these results are more relevant than results from Formula 4.1 in the proposed approach to determine the coverage of the insurance. The sum of the results is 100%, which means that if coverage of the insurance is divided by each decision percentage the coverage of each decision is obtained.

After this calculation, the influence of each decision of the transport process is known with a respect to the entire survey. These results are needed to calculate the coverage of the insurance.

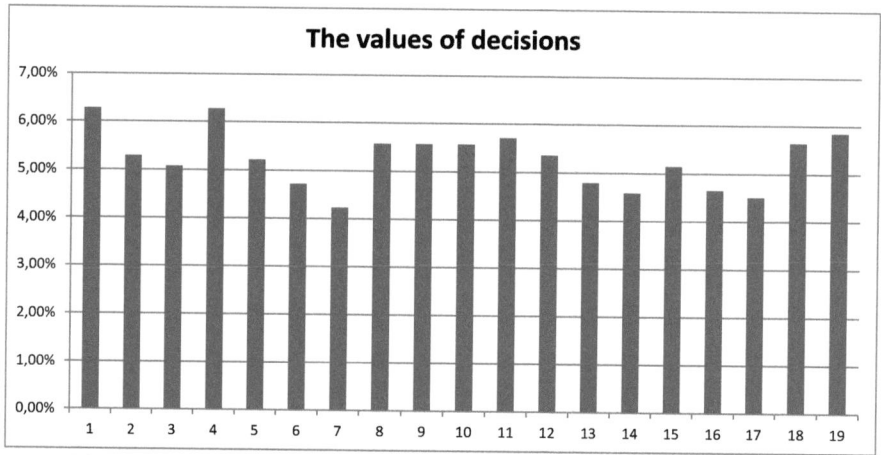

Figure 4.1: Graphical representations of applying Formula 4.2

As in the case of applying Formula 4.1 decision one and four are the most important and the same conclusions mentioned above apply.

4.3 A PROPOSED METHODOLOGY FOR TRANSPORT MANAGER INSURANCE

The proposed methodology for the insurance estimation is based on splitting all the decisions into three groups according to their importance. These groups were selected by using the results shown in Table 4.5. This methodology is applicable only for common goods and not for special goods such as animals, dangerous goods, food, and etc.

The group number one is comprised of the first eight decisions in Table 4.5. These decisions were chosen because they have a value higher than 5.5 %.

The group number two is comprised of another five decisions in the Table 4.5. These decisions have percent value of at least 5 % and at the most 5.5 %.

The last group includes six decisions and these decisions have at the most a 5 % percent value.

Group number one

This group includes the eight most important decisions in the whole transportation process. These are decisions, which a transport manager has to make before or during transportation process. These eight decisions make the core of the insurance proposal; these decisions have to be insured in all transportation process.

The decisions are:
- decision 1: ordering transport timely,
- decision 4: dates of transportation,
- decision 19: storage of goods during transportation,
- decision 11: neglect of professional car at order of transport,
- decision 18: filling the bill of lading,
- decision 8: responsibility for wrong decision,
- decision 9: late delivery,
- decision 10: correct handling with goods.

Group number two

This group is comprised of less important decisions but still important.
The decisions are:

- decision 12: responsibility for averting damage to the goods,
- decision 2: correct information about goods,
- decision 5: price,
- decision 15: not to inform the customer about destroyed goods,
- decision 3: correct vehicle.

Group number three

In the last group there are six decisions. These decisions are a kind of specialty decisions, because the transport manager has to make these decisions only in some special transportation processes or in special situations. The special transportation process means transportation of animal, foods, dangerous goods, expensive goods, etc. In these cases the transport manager must use a special vehicle; he/she needs better cover of his or her responsibility (reinsurance). In some cases some decision are dependent on the decision from groups one or two. Reinsurance means special insurance for less money, because transport manager has already a contract of insurance coverage with the same insurance company.

The decisions are:

- decision 13: insurance,
- decision 6: notice to the wrong customer's instructions about transportation,
- decision 16: neglecting of selling of goods, if goods can be the destroyed,
- decision 14: responsibility for another transport manager,
- decision 17: Non- application of his or her rights of the goods,
- decision 7: responsibility for the completion of the instructions.

4.3.1 COVERAGE OF THE KOOPERATIVA VIENNA INSURANCE GROUP OF COMPANY

This insurance company was used to explain the problem of insurance of the transport manager and then to explain the proposed solution to this problem.

Table 4.6 shows company Kooperativa Vienna Insurance Group insurance premiums for coverage of the transport manager responsibility at present.

Table 4.6: Insurance premiums for coverage of the transport manager responsibility

Coverage of the insurance (EUR)	Cost of the insurance (EUR)
16 596.96	232.36
33 193.92	331.94
66 387.84	564.30
99 581.76	829.85
132 775.68	995.82
165 969.59	**1 294.56**
199 163.51	1 593.31
232 357.43	1 858.86
265 551.35	1 958.44
298 745.27	2 091.22
331 939.19	2 389.96
497 908.78	3 253.00
663 878.38	4 315.21

Assume the highlighted numbers means the coverage of the insurance will be 165 969.59 € and cost of the insurance will be 1 294.56 €, then the coverages of each decision are shown in Table 4.7.

Tables 4.7: Coverage of each decision with coverage 165 969.59 €

Qn		
Question	Percent	Coverage of insurance company
1	6.27%	10 409.65 €
4	6.27%	10 409.65 €
19	5.85%	9 707.88 €
11	5.71%	9 473.95 €
18	5.64%	9 356.99 €
8	5.57%	9 240.03 €
9	5.57%	9 240.03 €
10	5.57%	9 240.03 €
SUM group 1	46.44%	77 078.20 €
12	5.36%	8 889.14 €
2	5.29%	8 772.18 €
5	5.21%	8 655.21 €
15	5.14%	8 538.25 €
3	5.07%	8 419.64 €
SUM group 2	26.07%	43 274.42 €
13	4.79%	7 953.44 €
6	4.72%	7 836.48 €
16	4.65%	7 719.52 €
14	4.58%	7 602.55 €
17	4.51%	7 485.59 €
7	4.23%	7 017.74 €
SUM group 3	27.48%	45 615.32 €
Σ	100%	165 969.59 €

The table shows how much money covers each kind of decision and each kind of group. If the coverage of the insurance is assumed to be 165 969.59 € and cost of the insurance 1 294.56 €; the coverage of the insurance for group one is 77 078.20 €, for group two is 43 274.42 € and for group three is 45 615.32 €.

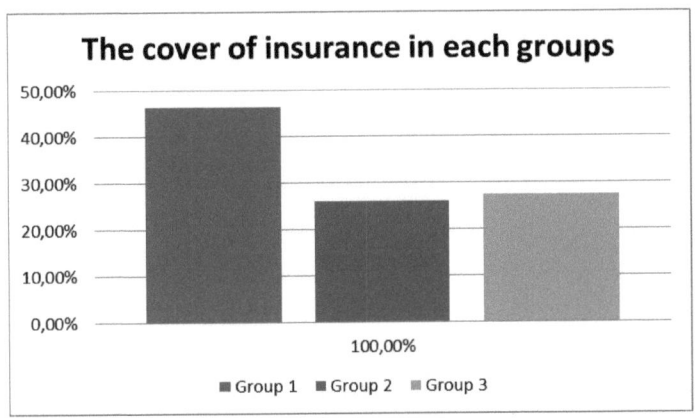

Figure 4.2: The graph of the coverage of insurance in each groups

Figure 4.2 shows that the group three has bigger coverage of insurance than group two. That is not good because in Chapter 4.3 there was found out, based on the survey, that group two is more important than group three. The coverage of group three is needed only in special cases.

4.3.2 OPTIMIZE THE COVERAGE OF COMPANY KOOPERATIVA VIENNA INSURANCE GROUP

It is known that the decisions in group three are for special cases of transportation or in special situations. That means there are two options. First, make transfer of money from group three to other groups or second, the transport manager can use this money to develop his/ her company.

4.3.2.1   MOVING MONEY FROM GROUP THREE TO OTHER GROUPS TO IMPROVE INSURANCE COVERAGE

If money was moved from group three to other groups to cover the insurance, the money has to be reallocated to each of the decisions in groups one and two proportionally. Also the percentage value has to be recalculated. Group three has 27.48 % and 45 615.32 €. The calculation of new percentages in groups one and two without group three has to be computed. These percentages are calculated as:

$$\frac{27.48\%}{13} = 0.02113846$$

The number thirteen is the number of the sum decisions in groups one and two. The result 0.02113846 is the number that has to be added to each decision. Then coverage will be calculated again to find out how much money covers for which decisions and which groups.

Table 4.8: Coverage for each question without group three

Question	Percent	Coverage of insurance company
Qn		
1	8.39%	13 917.99 €
4	8.39%	13 917.99 €
19	7.96%	13 216.22 €
11	7.82%	12 982.29 €
18	7.75%	12 865.33 €
8	7.68%	12 748.37 €
9	7.68%	12 748.37 €
10	7.68%	12 748.37 €
SUM	63.35%	105 144.93 €
12	7.47%	12 397.48 €
2	7.40%	12 280.52 €
5	7.33%	12 163.56 €
15	7.26%	12 046.59 €
3	7.19%	11 927.98 €
SUM	36.65%	60 816.17 €
Σ	100.00%	165 969.59 €

Assume again that the coverage of the insurance is 165 969.59 € and cost of the insurance is 1 294.56 €; the coverage of the insurance for group 1 is 105 144.93 € and for group 2 is 60 816.17 €

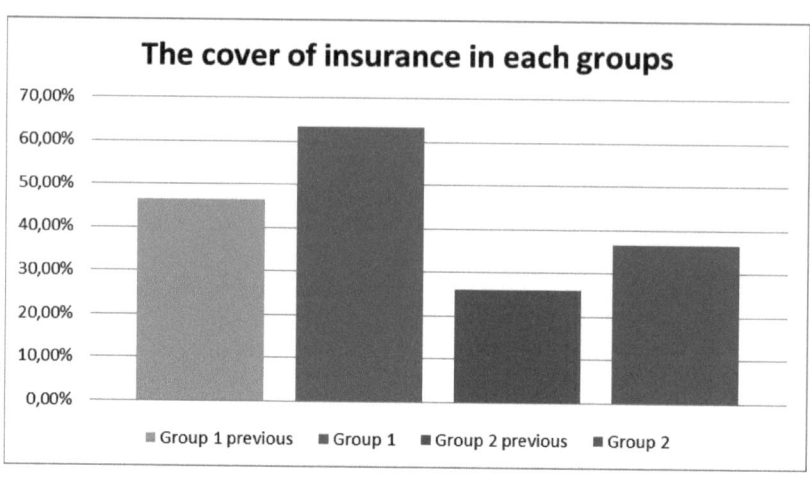

Figure 4.3 Values of all decisions in a survey

Figure 4.3 shows that after this calculation coverage of the insurance is bigger for each group. The coverage of group one was 77 078.20 € and now it is 105 144.93 €; the difference between these numbers is 28 066.73 €. The coverage of group two was 43 274.42 € and now it is 60 816.13 €; the difference between these numbers is 17 541.71 €. This means the transport manager will have better coverage of his responsibility during the transportation process, but if he/she wanted reinsurance of his or her responsibility he/she will have to pay for this reinsurance extra money.

4.3.2.2 THE WAY OF HOW THE MONEY FROM GROUP THREE SHOULD BE USED

The best way to protect the transport manager's responsibility and save money at the same time is by developing a special kind of insurance which will focus on group one and group two and group three will be like bonus insurance.

Tables 4.9: Coverage and cost of the insurance for each question

Qn Question	Percent	Coverage of insurance company	Cost of the insurance
1	6,27%	10 409,65 €	81,20 €
4	6,27%	10 409,65 €	81,20 €
19	5,85%	9 707,88 €	75,72 €
11	5,71%	9 473,95 €	73,90 €
18	5,64%	9 356,99 €	72,98 €
8	5,57%	9 240,03 €	72,07 €
9	5,57%	9 240,03 €	72,07 €
10	5,57%	9 240,03 €	72,07 €
SUM	46,44%	77 078,20 €	601,21 €
12	5,36%	8 889,14 €	69,34 €
2	5,29%	8 772,18 €	68,42 €
5	5,21%	8 655,21 €	67,51 €
15	5,14%	8 538,25 €	66,60 €
3	5,07%	8 419,64 €	65,67 €
SUM	26,07%	43 274,42 €	337,54 €
13	4,79%	7 953,44 €	62,04 €
6	4,72%	7 836,48 €	61,12 €
16	4,65%	7 719,52 €	60,21 €
14	4,58%	7 602,55 €	59,30 €
17	4,51%	7 485,59 €	58,39 €
7	4,23%	7 017,74 €	54,74 €
SUM	27,48%	45 615,32 €	355,80 €
Σ	100%	165 969,59 €	1 294,56 €

In Table 4.9 it can be seen that main coverage of the responsibility is the sum of group one and group two.

$$77\ 078.20\ € + 43\ 274.42\ € = \mathbf{120\ 352.62\ €}$$

where the sum is total coverage for decisions from groups one and two. The decisions in the group three are only for special cases of transportation or in special situations that means the transport manager needs coverage of group three just in special cases.

A solution is that the transport manager will pay **1 294.56 €** per year because he/she cannot anticipate what kind of coverage will be needed or what happens during the transport process, but if he/ she will not use the coverage of group three during the whole year the bank give him/her money back from group three coverage which is **355.80 €.**

CHAPTER 5: CONCLUSION

The goal of my thesis was to compare how to start business in the transport management to the U.S. and the European Union. The U.S. and the EU have uniform general business conditions. In the U.S. there are no specialized conditions. The only thing the applicant has to do to start a business is to go to the appropriate office and sign up for the business. The conditions within the EU are stricter, unlike the U.S. The applicant has to be in age of at least 18 years, legal capacity and integrity.

The professional conditions of the transport manager in the U.S. are better. The transport manager has to have compulsory insurance "BOND" which is a type of insurance, age over 21 and a broker´s license, which is obtained after successfully passing a test. The EU has not unified professional conditions for business. Some countries have easier conditions (Czech Republic) for transport managers and some countries have stricter ones (Austria). The EU needs to unify professional conditions in the field of the transport management for all the countries in EU, because the competitive environment is distorted.

Another section of this thesis is devoted to the responsibility of the transport manager. The thesis shows all the decisions for which the transport manager is responsible during the transportation process and what are his or her responsibilities when working under transport manager´s conditions. Based on these decisions, a diagram that shows the decisions in sequence according to their importance is presented. For a better understanding of the transport manager's responsibilities a fictitious transportation process was developed to simulate what should happen if the transport manager made the wrong decisions.

To understand the values of the decisions a questionnaire was created with individual decisions. The survey was sent to selected logistics and transport manager companies to fill. Then transport companies added a value for each decision. A total of 48 answers from transport companies and from transport managers were prepared.

After surveys from the transport companies where received; calculation started to identify which decisions are less important and which are more important. Two formulas were proposed to find the value of a decision. The first formula is just a weighted average of the responses for each answer. The second formula calculates the value with respect to the entire questionnaire. Results show that the most important decision of the questionnaire is decision is number one ordering transport timely.

After finding that some decisions are less important than the others, these decisions were divided into the three groups. Group one is the most important decisions that the transport manager has to have insured. The second group is the less important decisions, but still it is better to be insured. And the last category (number three) contains decisions, which the transport manager has to decide only in special cases and situations.

Then, a tariff was chosen from a table of the insurance company Kooperativa and insurance premiums for each group of decisions were calculated. The group three has larger financial coverage than the group two and thus had to be changed since the group two is more important than the group three.

The transport manager had two options. The transport manager can move money from group three to the other groups, then the coverage of insurance will be higher, or he/she does not move money to the other groups and the transport manager can save this money and invest it.

The proposed solution is that the transport manager will pay the full price of insurance coverage per year as it is specified in tables of tariff of the insurance companies. The reason is that he/she cannot anticipate what coverage will be needed or what can happen during the transportation process. If he/she does not use coverage from group three during the year, the bank gives him or her money back from the coverage of group three or the transport manager will pay less the following year.

REFERENCES

[1] http://www.koop.sk/index.cfm?module=lomtec&page=SearchResults

[2] Commercial Code no. 513/1991 collection of Laws from 5th November 1991

[3] Directive 2006/123/EC of the European Parliament and of the council of 12 December 2006 on services in the internal market

[4] 136/2010 Collection of Laws for Slovakia and above- mentioned European Directive about services in the internal market. (http://www.zbierka.sk/zz/predpisy/default.aspx?PredpisID=209600&FileName=z z2010-00136-0209600&Rocnik=2010)

[5] Government Resolution no. 140/2000 Collection of Law (http://www.sagit.cz/pages/sbirkatxt.asp?zdroj=sb00140&cd=76&typ=r)

[6] Customs Act (http://www.vyvlastnenie.sk/predpisy/colny-zakon/)

[7] Act no. 455/1991 Collection of Laws on business activities. (http://www.zbierka.sk/zz/predpisy/default.aspx?PredpisID=11370&FileName=91-z455&Rocnik=1991)

[8] http://www.telecom.gov.sk/index/index.php?ids=1

[9] http://www.ok.gov/opm/jfd/t-specs/t23.htm; http://www.wisegeek.com/what-does-a-transportation-manager-do.htm

[10] http://smallbusiness.chron.com/differences-between-llc-sole-proprietor-s-corp-4372.html

[11] http://www.cbp.gov/xp/cgov/trade/trade_programs/broker/brokers.xml

[12]http://www.google.com/url?sa=t&rct=j&q=&esrc=s&source=web&cd=1&sqi=2 &ved=0CHcQFjAA&url=http%3A%2F%2Fwww.cbp.gov%2Flinkhandler%2Fcg ov%2Ftrade%2Ftrade_programs%2Fbonds%2Fqa_bonds.ctt%2Fq_and_a_bonds. doc&ei=qzdqT9yBOung2QXUr8TsCA&usg=AFQjCNG719Kf9DQo2TodUPFd W5h1Mqw_eg&sig2=otLqkB5K90HCMniq_BLy9Q

[13] http://www.zlz.sk/sk/informacie-o-zvaeze/zasielateske-podmienky/48-
zasielateske-podmienky.html

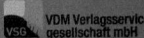